This book belongs to:

....

....

1 dragon

1 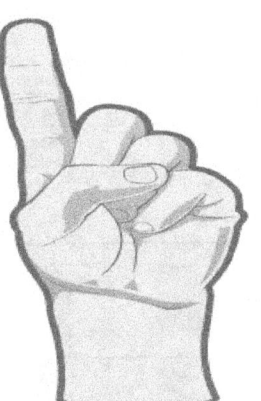 one

1

1

one one one

one

1  one

One apple

1 1 1

1

one one one

one

2 polar bears

2 two

2 2 2

2

two two two

two

2 two

Two apples

2 2 2

2

two two two

two

3 birds

3 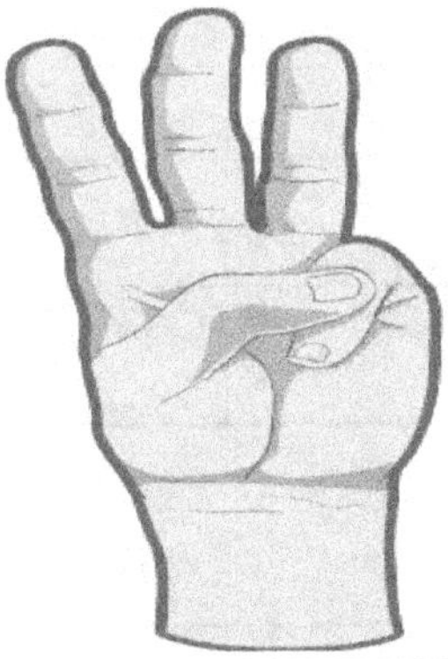 three

3 3 3

3

three three

three

3  three

Three apples

4 dogs

4   four

four four four
four

4   four

Four apples

4 4 4

4

four four four

four

5 cup cakes

5 five

5 5 5

5

five five five

five

5 five

5 5 5

5

five five five

five

6 fish

6 six

6 6 6

6

six six six

six

6 six

7 hats

7 seven

7 7 7

7

seven seven

seven

7 seven

8 cars

8 eight

8

8

eight eight

eight

8 eight

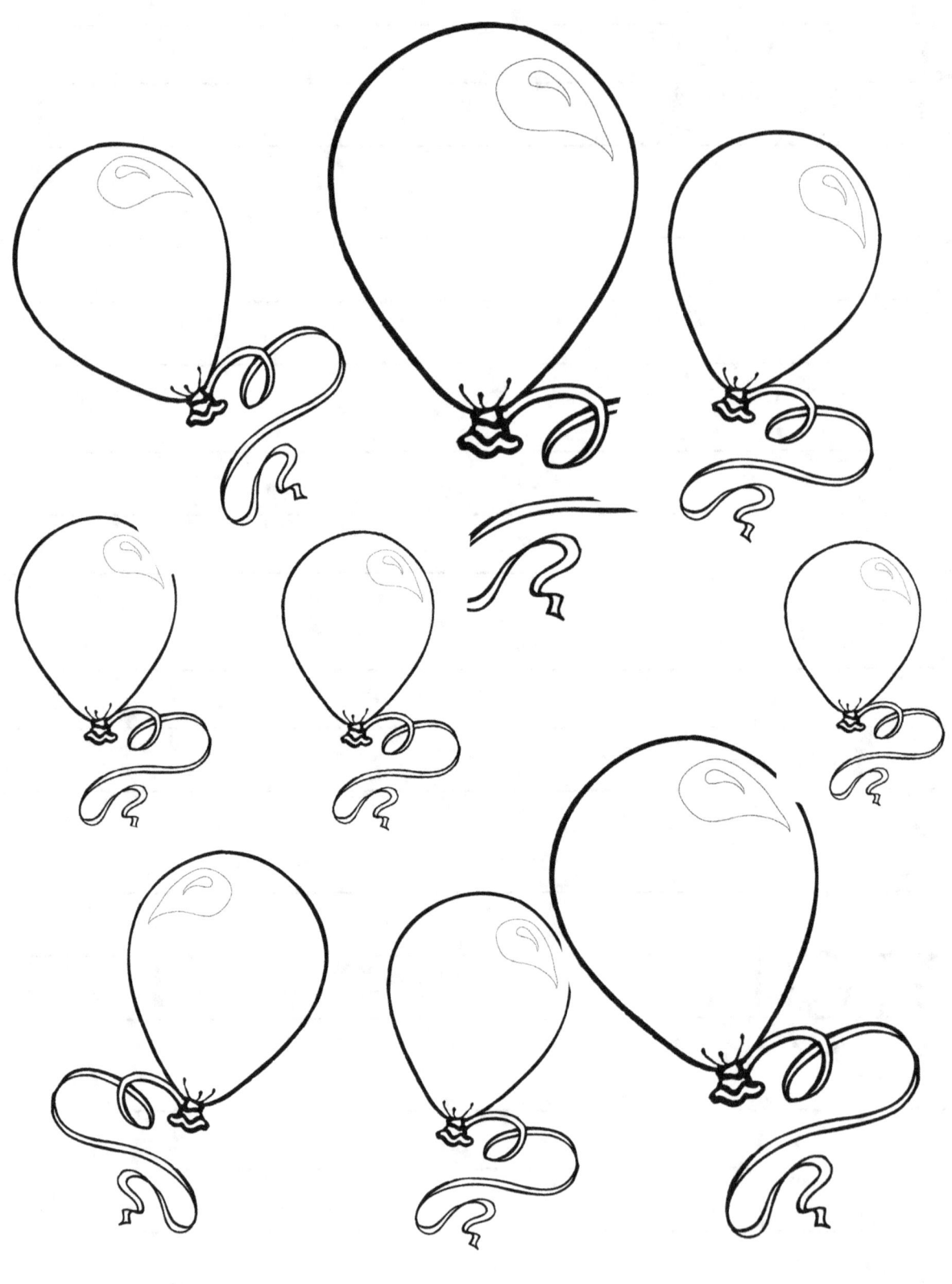

9 balloons

9 nine

9 9 9

9

nine nine nine

nine

9 nine

10 stars

10 ten

10 10 10

10

ten ten ten

ten

10 ten

10 10 10

10

ten

ten

Draw a line from the number to the corresponding number of items:

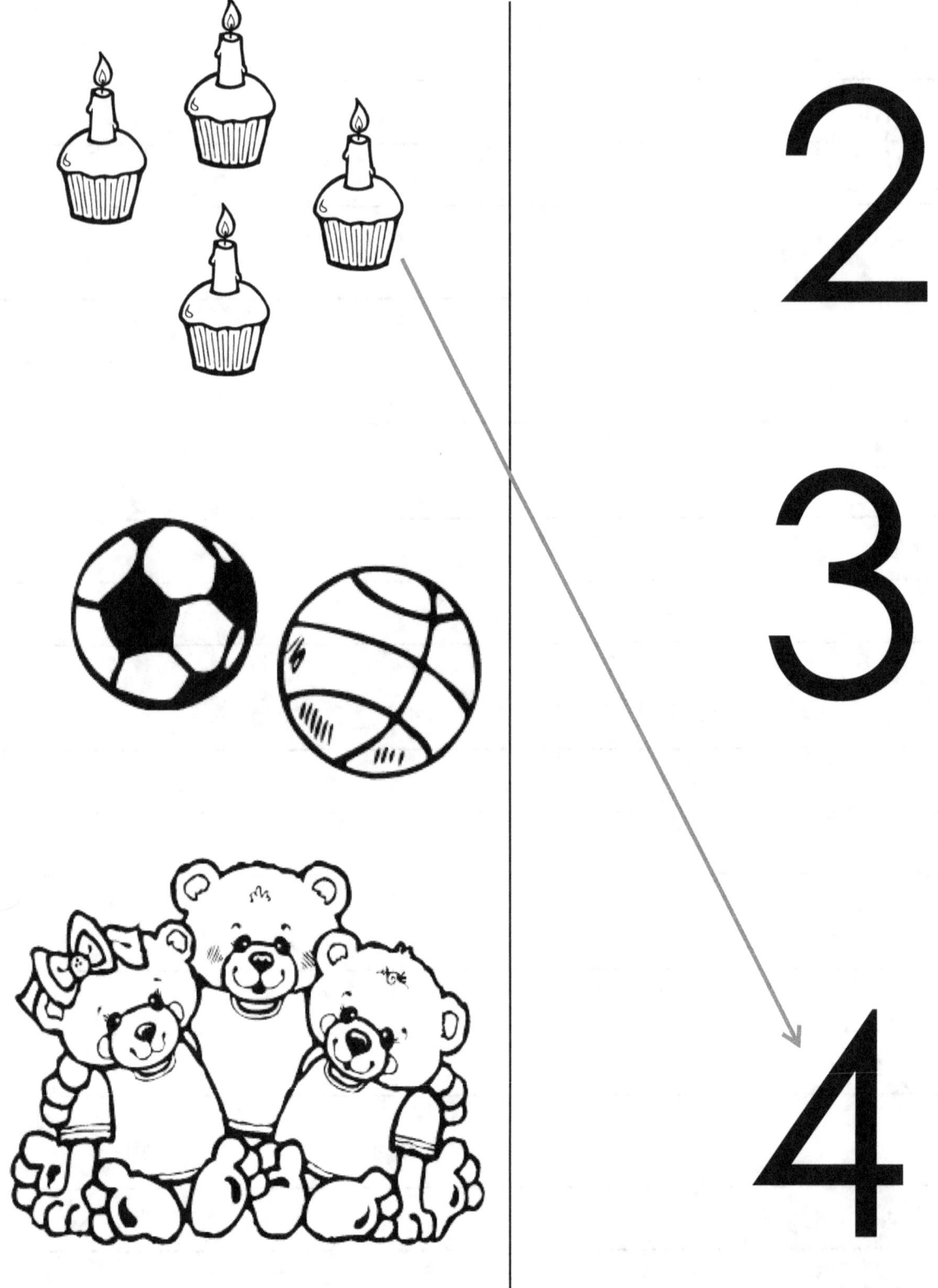

Draw a line from the number to the corresponding number of items:

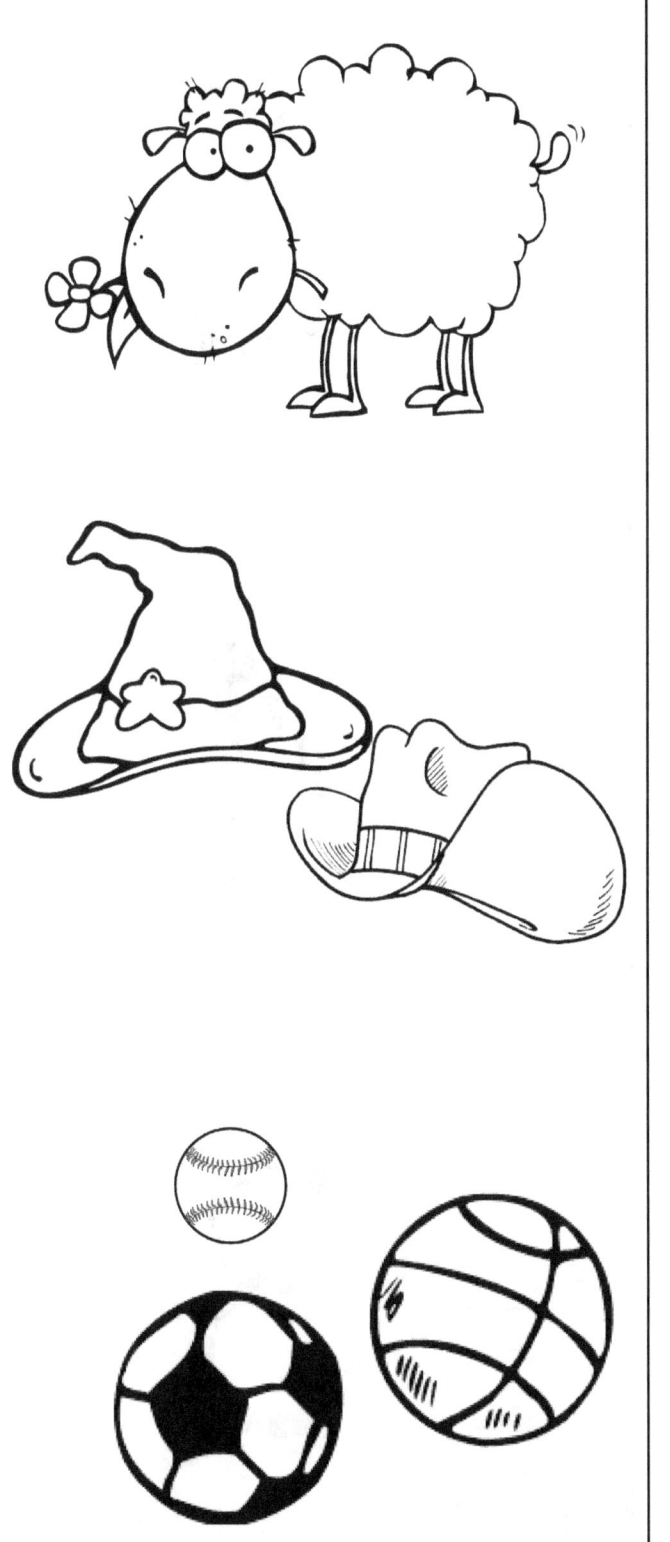

2

1

3

Draw a line from the number to the corresponding number of items:

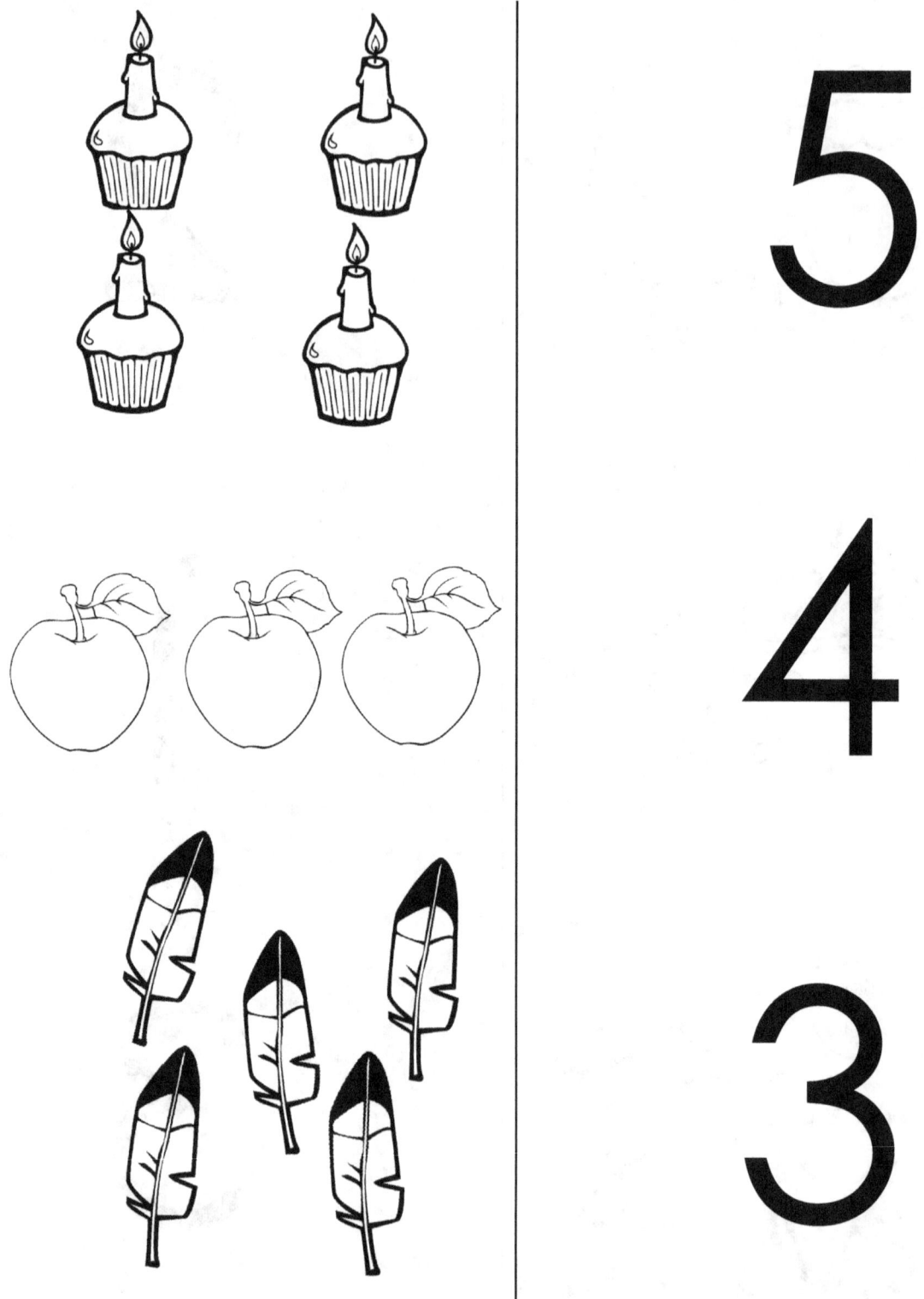

Draw a line from the number to the corresponding number of items:

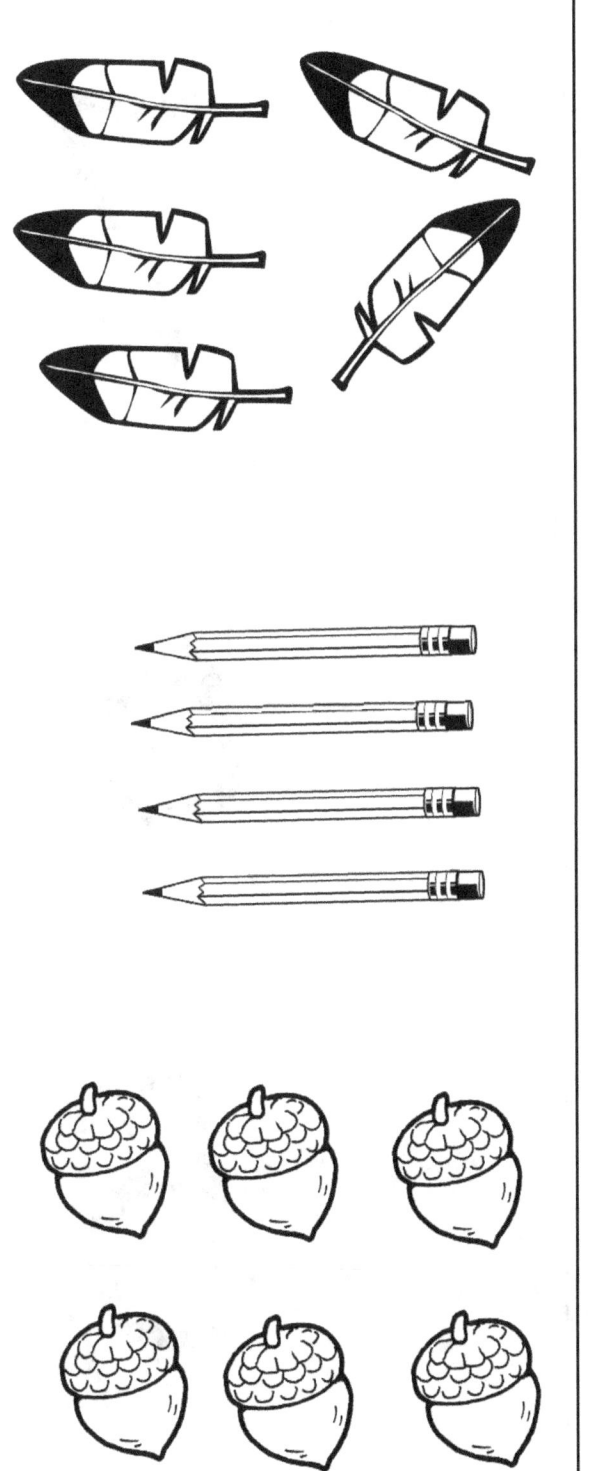

Draw a line from the number to the corresponding number of items:

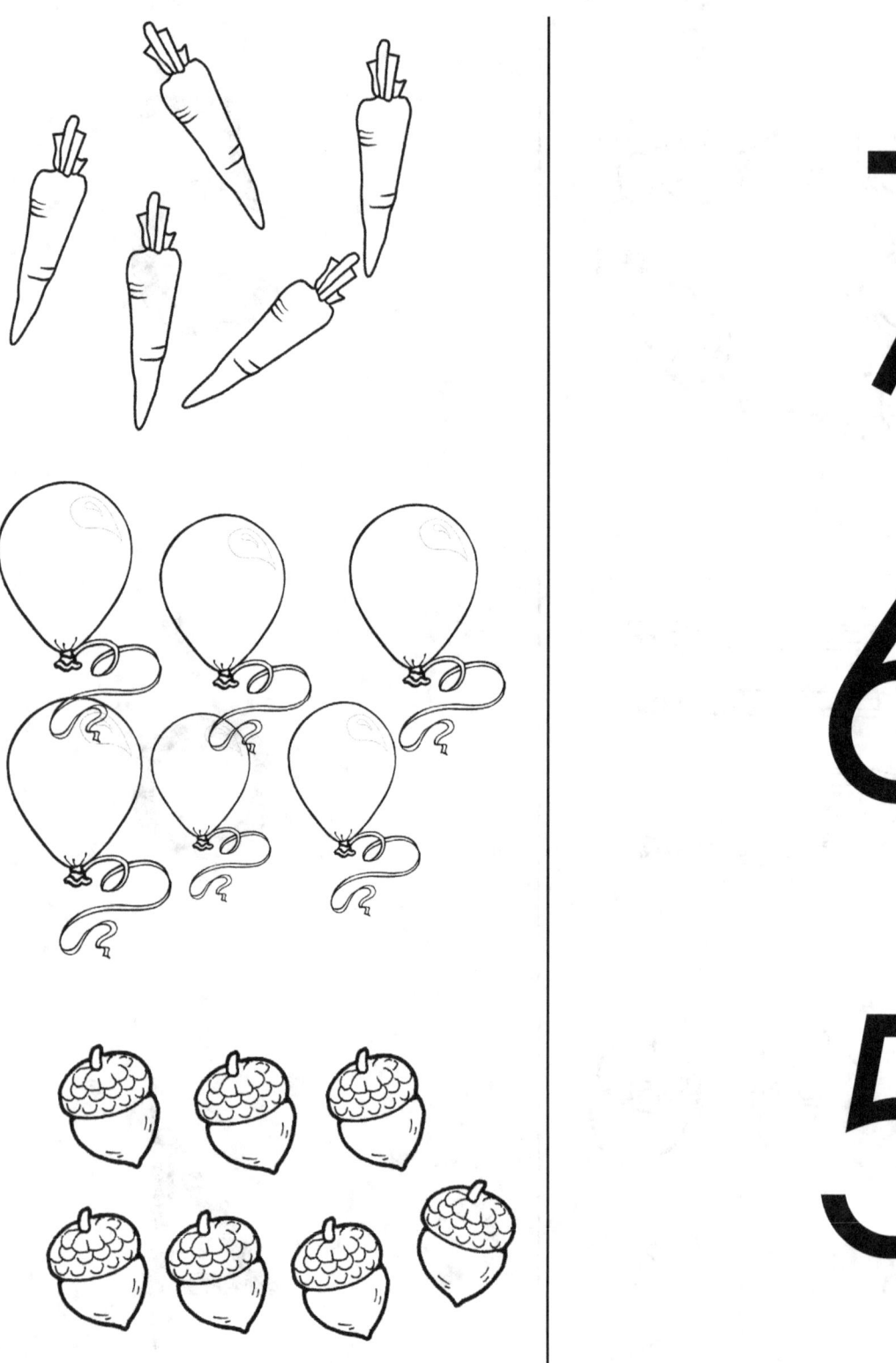

Draw a line from the number to the corresponding number of items:

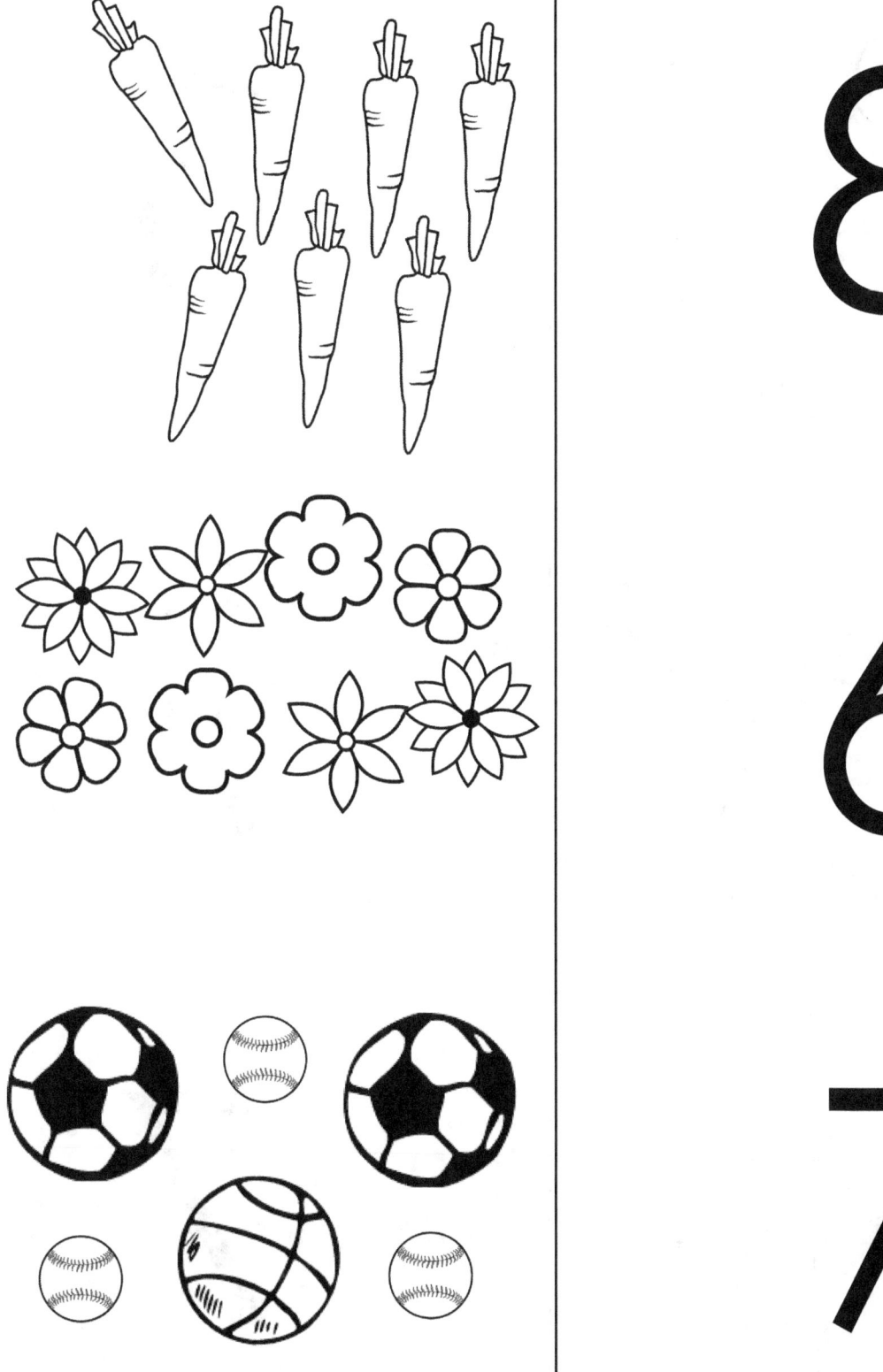

Draw a line from the number to the corresponding number of items:

Draw a line from the number to the corresponding number of items:

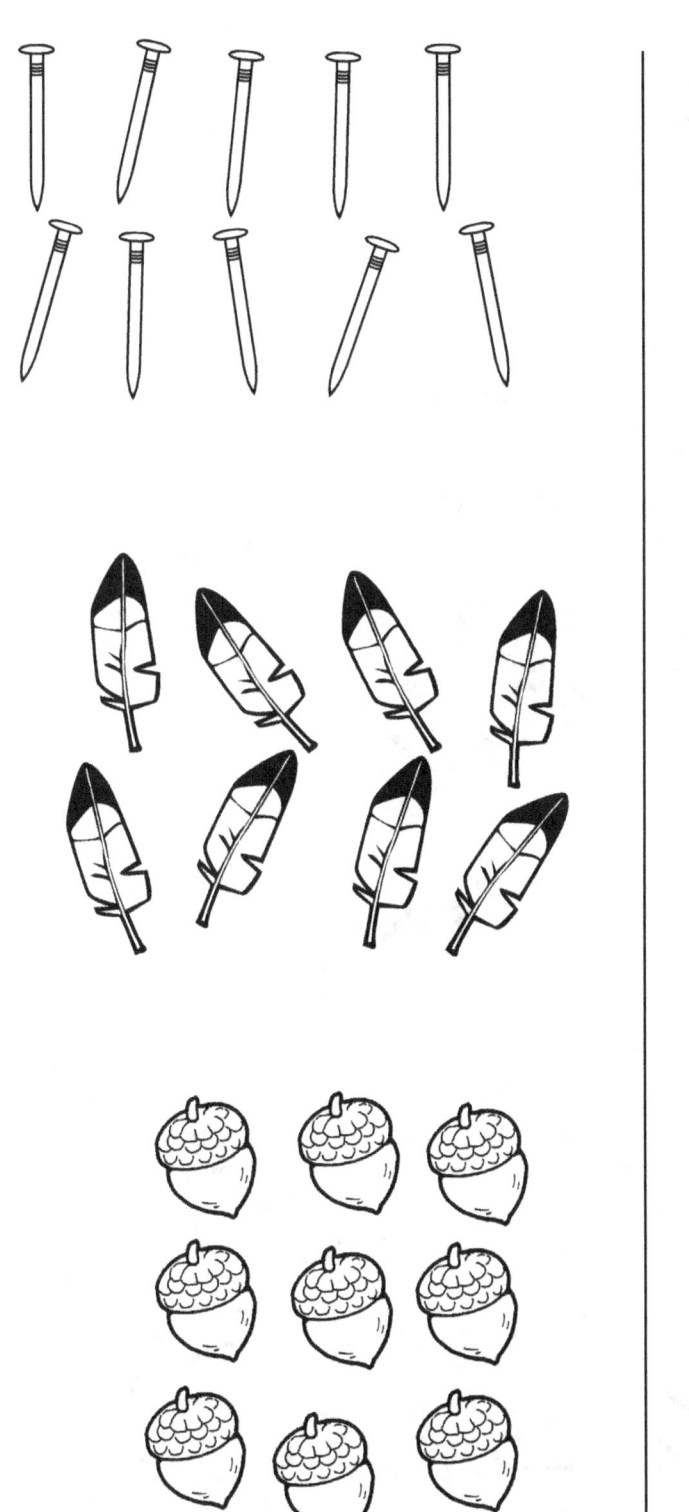

Count the animals and circle the right number:

1 2 3

Count the animals and circle the right number:

1 2 3

Count the animals and circle the right number:

1 2 3

Count the animals and circle the right number:

Count the animals and circle the right number:

4 5 6

Count the animals and circle the right number:

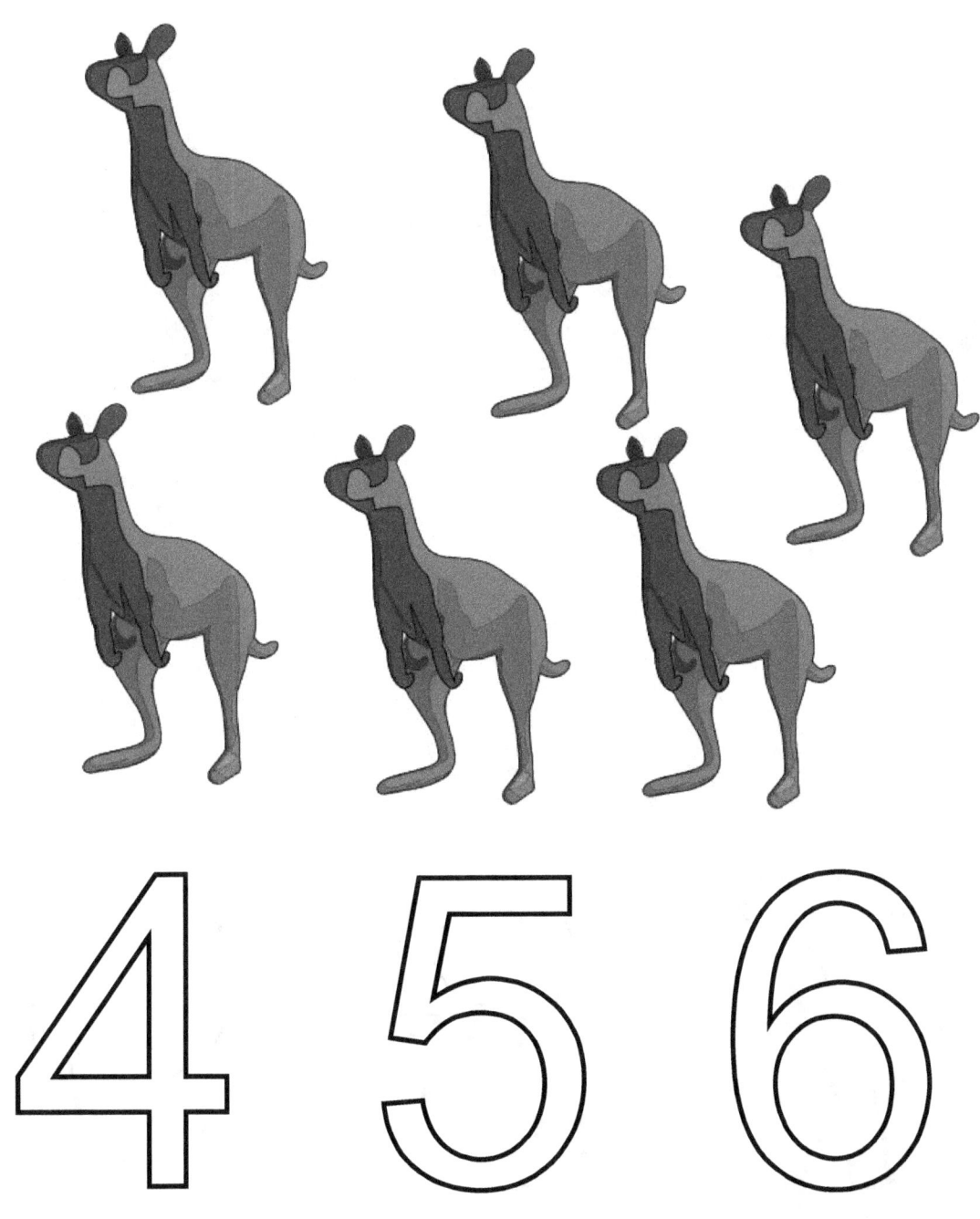

4 5 6

Count the animals and circle the right number:

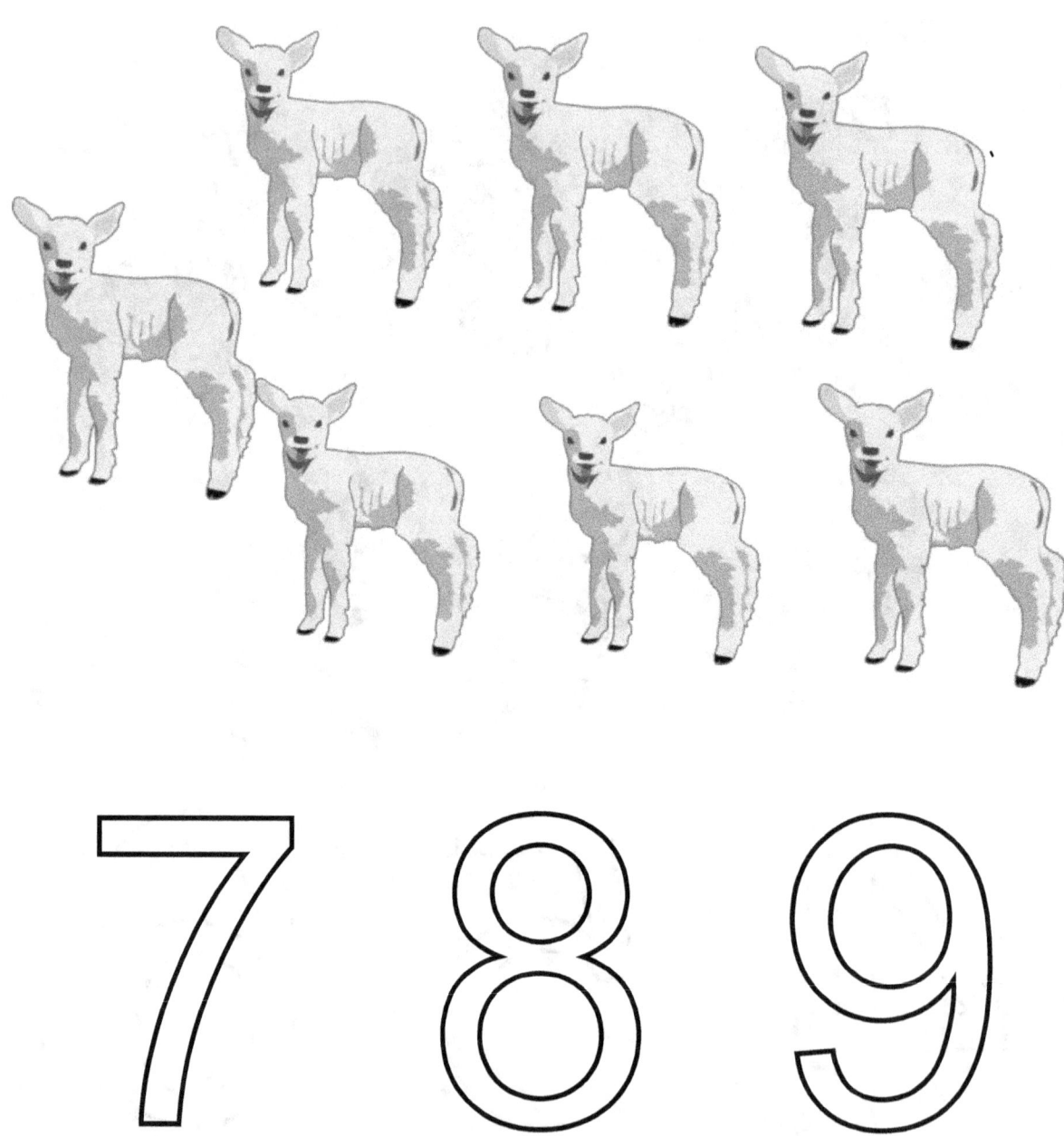

7 8 9

Count the animals and circle the right number:

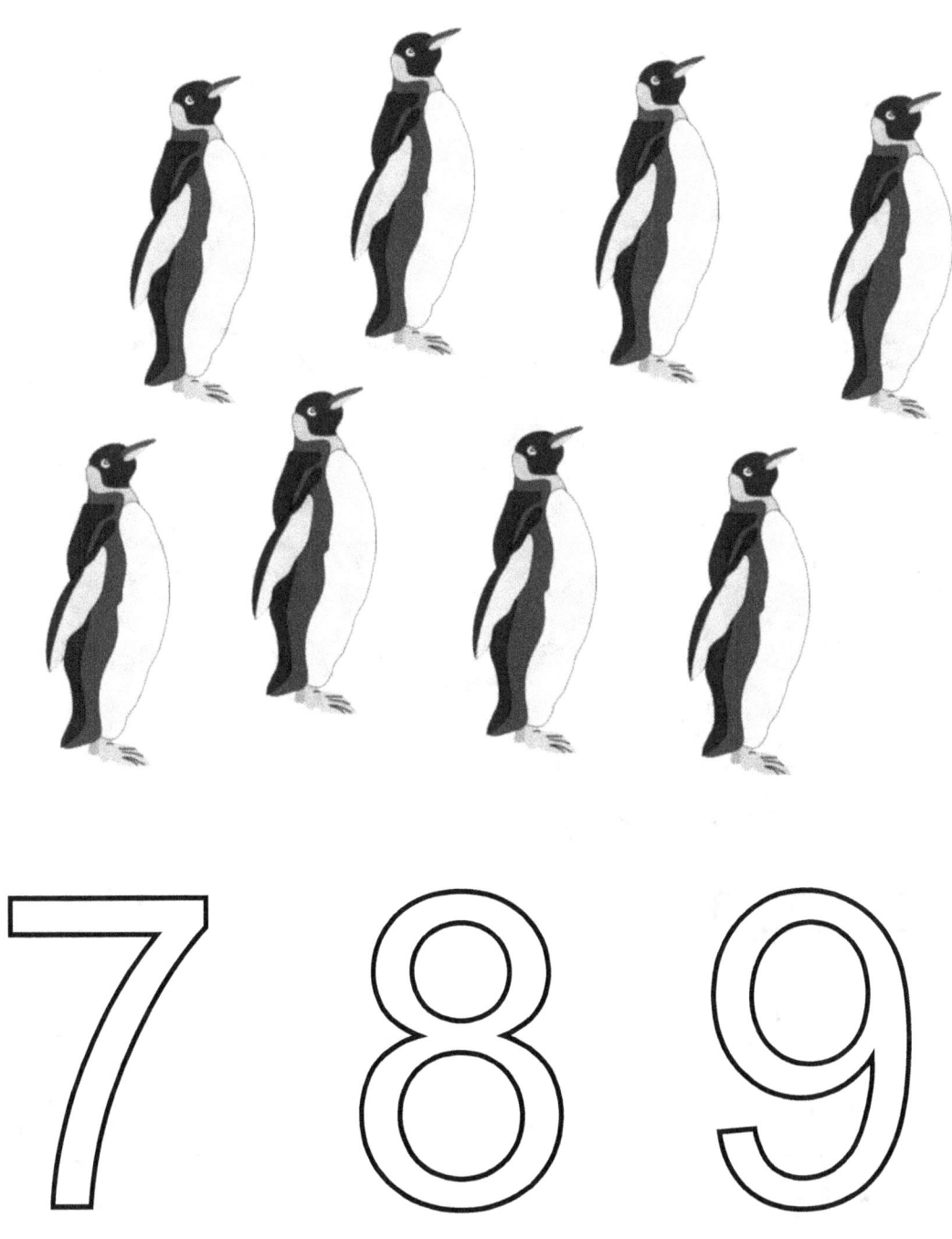

7 8 9

Count the animals and circle the right number:

Count the animals and circle the right number:

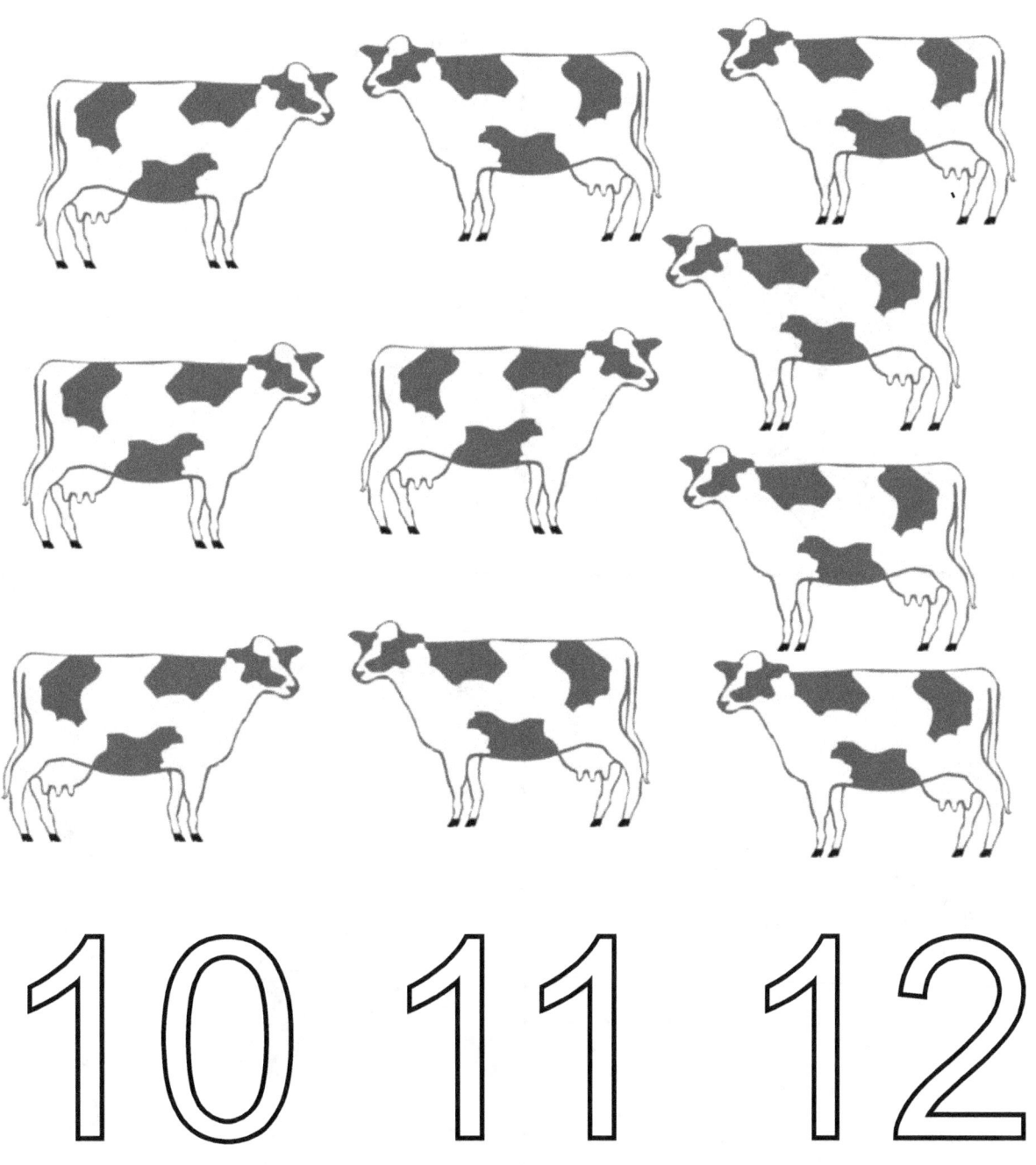

10 11 12

Count the animals and circle the right number:

How many _____?	How many _____?
♥♥♥	★★★★★
How many _____?	**How many _____?**
♪ ♪	💧 💧

Count the animals and circle the right number:

How many _____?

♥ ♥ ♥

How many _____?

★

How many _____?

♪ ♪
♪ ♪

How many _____?

💧 💧 💧
💧 💧 💧
💧 💧 💧

Count the animals and circle the right number:

Count the animals and circle the right number:

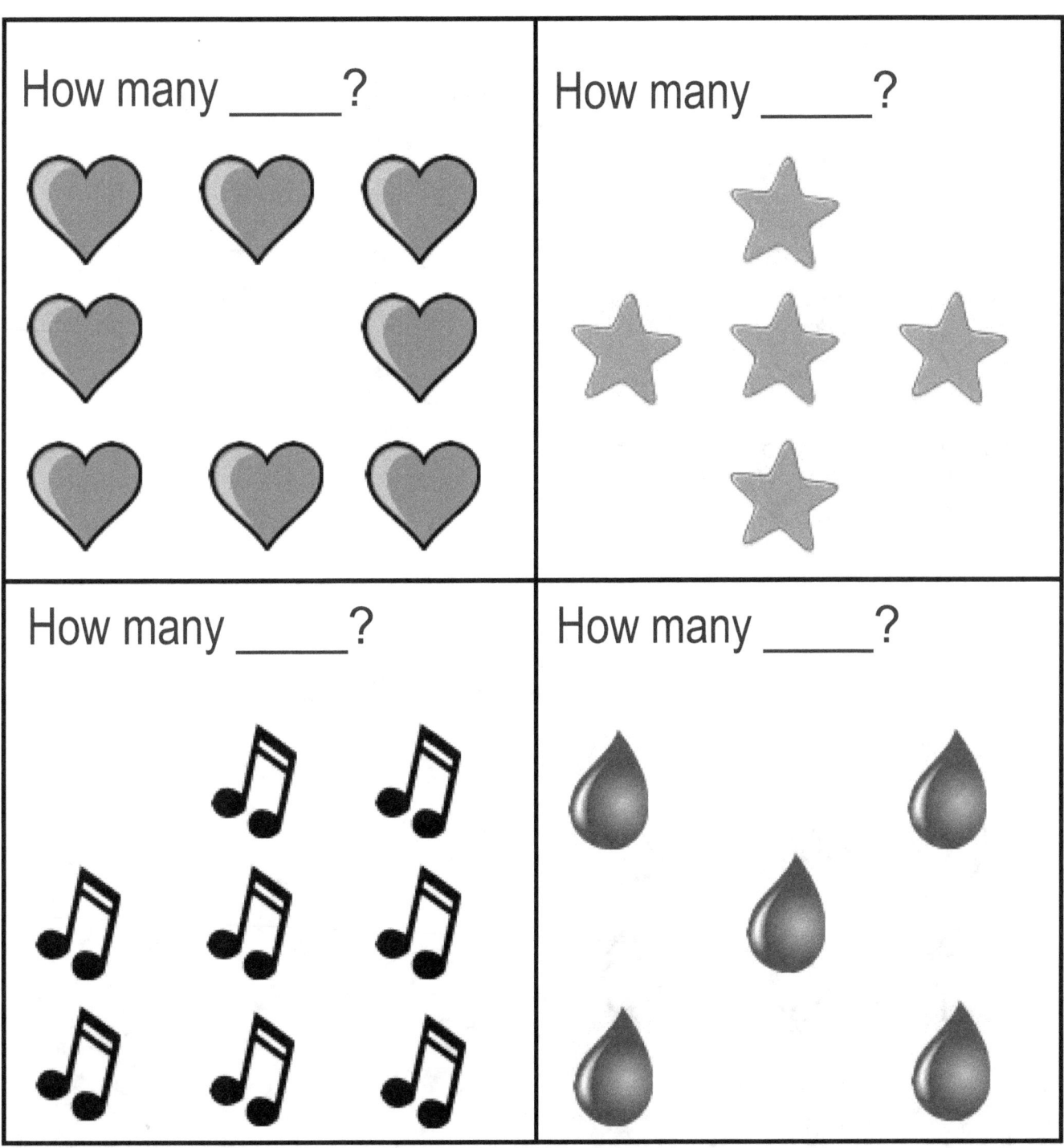

Count the animals and circle the right number:

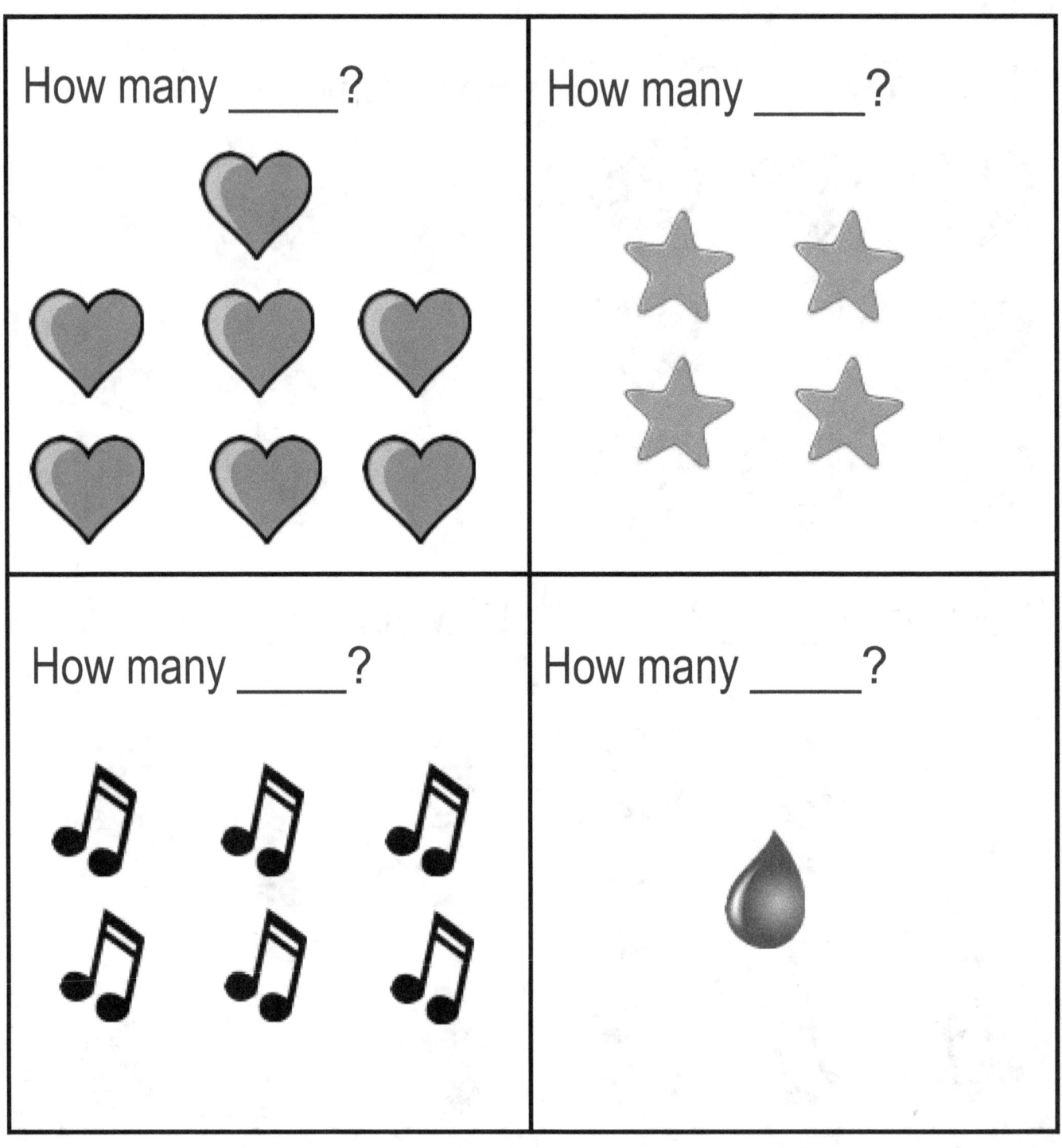

1 2 3 4 5 6 7 8 9 10

Count and add the items in each row.

# = _____

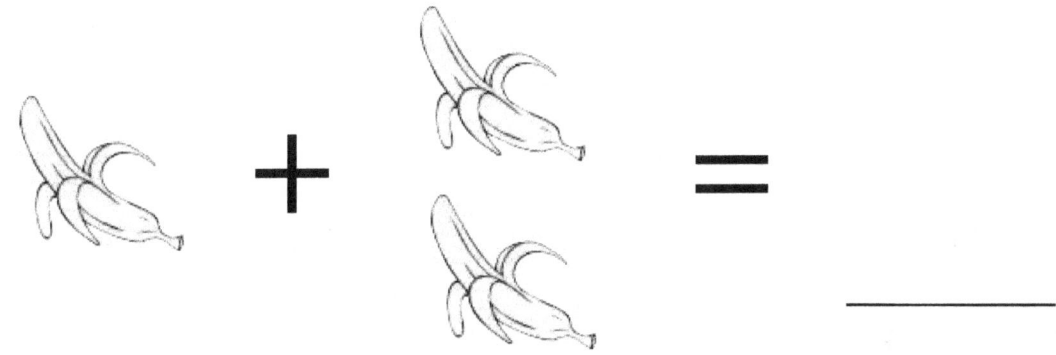# = _____

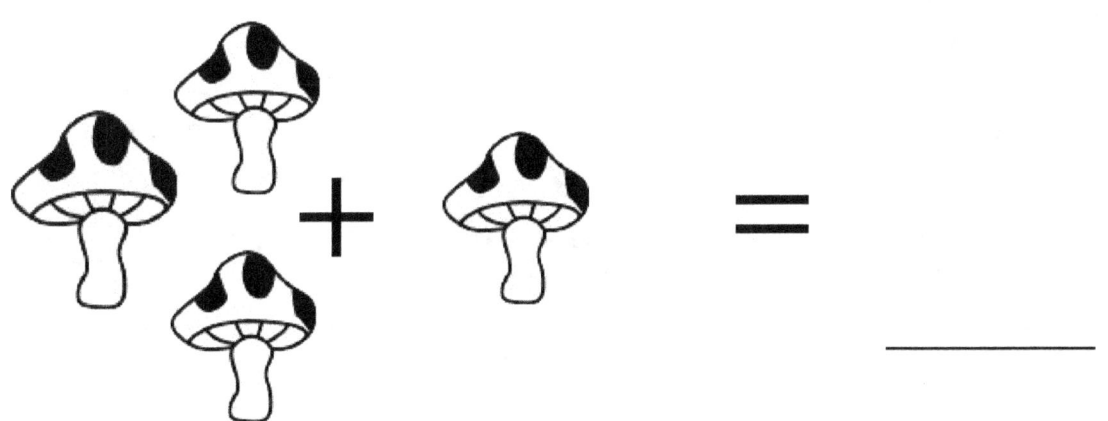# = _____

= _____

1 2 3 4 5 6 7 8 9 10

Count and add the items in each row.

Example

$\underline{3} + \underline{4} = \underline{7}$

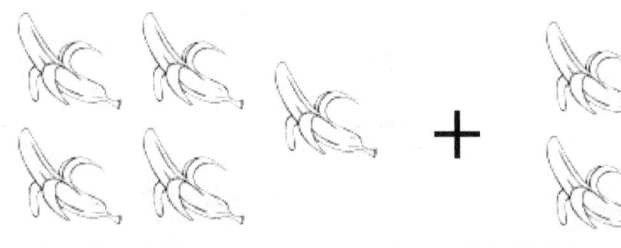

___ + ___ = ___

___ + ___ = ___

___ + ___ = ___

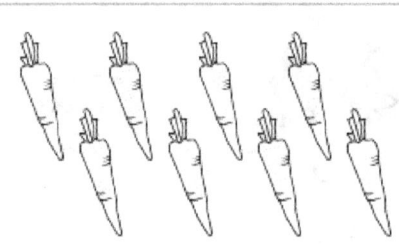

___ + ___ = ___

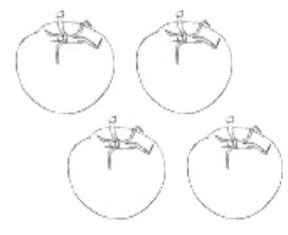

___ + ___ = ___

1 2 3 4 5 6 7 8 9 10

Count and subtract

= _____

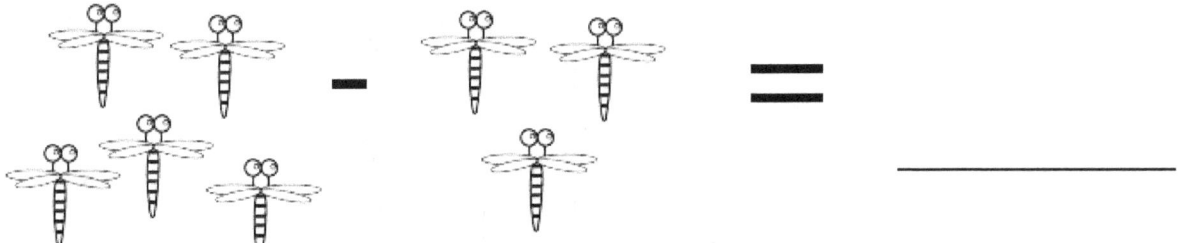
= _____

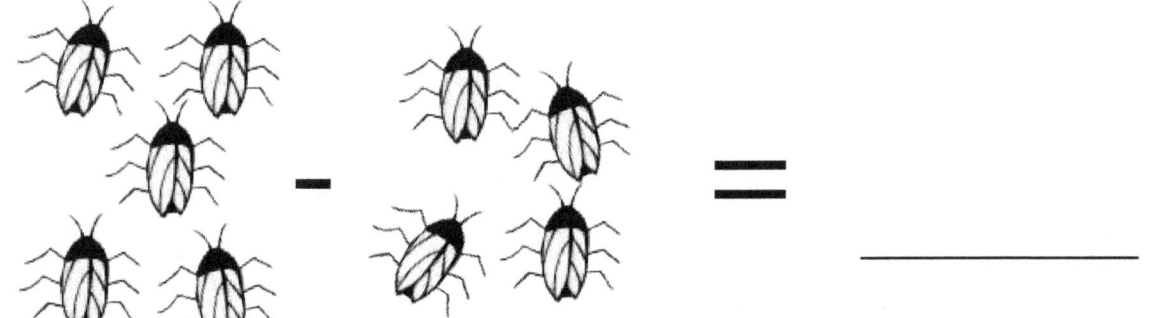
= _____

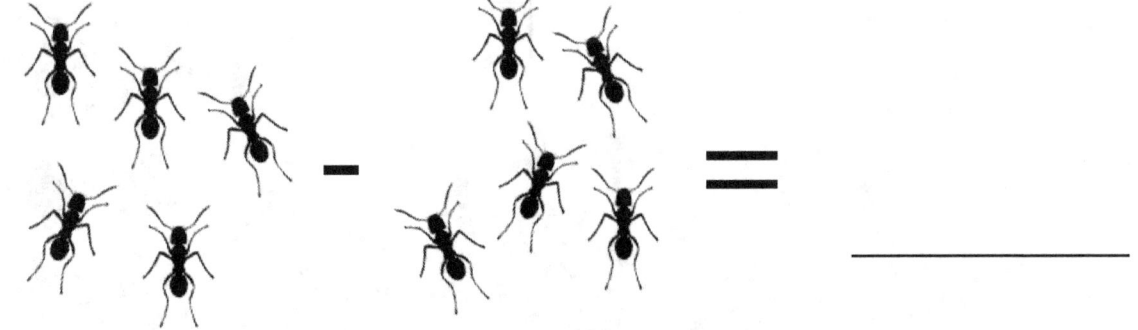
= _____

www.ingramcontent.com/pod-product-compliance
Lightning Source LLC
Chambersburg PA
CBHW080906220526
45466CB00011BA/3481